100000 Whys for Toddlers

Fun Questions and Answers About Animals Weather, and the Body for Preschoolers

Table of Contents

Little Hands, Big Questions

A happy book of big little wonders

Let's look, point, smile, and ask why!

Copy Me!
Can you point to your favorite picture?

Hello, Little Explorer!

This book is full of happy questions about your world.

Look at the pictures. Point to what you see. Then talk about them together.

Copy Me!
Can you wave hello like our little explorer?

How to Use This Book

1. Look

Find the big picture.

2. Ask

Say the question out loud.

3. Play

Try the fun action!

Copy Me!

Can you point to Step 1?

What Will We Learn About?

Animals

My Body

Sky & Weather

Toys & Play

Yummy Food

Nature

We will look, wonder, and smile on every page.

Copy Me!

Which picture do you like first?

Ready to Begin?

Turn the page and let's ask our first big question!

Look. Point. Ask. Play.

Copy Me!

Can you clap
your hands and say,
'Let's go!'

Part 1
Animals & Little Creatures

Let's meet furry, feathery, and fluffy friends!

Copy Me!

Can you make your favorite animal face?

Why do dogs wag their tails?

Dogs wag their tails to show how they feel.
A happy dog may wag to say hello!

Copy Me!

Can you wag like a happy puppy?

Why do cats purr like little motors?

Cats purr when they feel cozy, calm, or safe.
Their soft purr can mean they feel good.

Copy Me!

Can you curl up like a cozy cat?

Why can birds fly up high in the sky?

Birds have wings and light feathers to help them fly. Their bodies are made for soaring high.

Copy Me!

Can you flap your arms like a bird?

Why do ducks love splashing in puddles?

Ducks love water fun, and splashing helps keep their feathers clean.
Their feathers are made for wet days.

Copy Me!

Can you waddle like a duck?

Why do baby cows drink milk from their moms?

Baby cows drink milk from their moms to grow big, strong, and healthy. Milk gives them what they need.

Copy Me!

Can you say moo like a little calf?

Why do bunnies wiggle their noses?

Bunnies wiggle their noses to sniff the world around them. Their little noses help them explore.

Copy Me!

Can you wiggle your nose like a bunny?

Why do bees visit flowers?

Bees visit flowers to sip sweet nectar.
As they move, they help flowers make new seeds.

Copy Me!

Can you buzz like a tiny bee?

Why do fish swish their tails in the water?

Fish swish their tails to swim through the water. Their tails help push them along.

Copy Me!

Can you swish side to side like a fish?

Why do turtles hide in their shells?

Turtles tuck into their shells to feel safe and protected. Their shells are like little homes.

Copy Me!

Can you curl up small like a turtle?

Part 2
My Body & How I Move

Let's learn how our bodies rest, grow, and move every day!

Copy Me!

Can you point to your eyes, hands, and feet?

Why do we close our eyes when we sleep?

We close our eyes when we sleep so our body and brain can rest.
Sleep helps us wake up ready for a new day.

Copy Me!

Can you close your eyes and pretend to sleep?

Why do we laugh when someone tickles us?

Tickles can feel silly and surprising.
That funny feeling can make us laugh.

Copy Me!

Can you make your biggest happy laugh face?

Why do we need to wash our hands with soap?

Soap and water help wash away tiny germs and dirt.
Clean hands help keep us healthy.

Copy Me!

Can you rub your hands together like you are washing them?

Why do we wear big coats when it is cold outside?

A warm coat helps keep our body heat close to us.
It helps us stay cozy when the air feels cold.

Copy Me!

Can you give yourself a cozy hug like a warm coat?

Why do we grow taller every single year?

As we grow, our bones and muscles get bigger and stronger.
That helps us grow taller year by year.

Copy Me!

Can you stretch up high and tall like you are growing?

Why do our feet help us walk and run?

Our feet help hold us up and move us along.
They help us walk, run, and play.

Copy Me!

Can you march your feet like a little walker?

Why do our hands help us hold and clap?

Our hands help us grab, hold, wave, and clap.
They help us do so many fun things.

Copy Me!

Can you clap your hands for the book?

Why does our nose help us smell?

Our nose helps us smell the world around us.
It can help us notice sweet, yummy, or fresh smells.

Copy Me!

Can you sniff like you smell a flower?

Why do our ears help us hear?

Our ears help us hear sounds around us.
They help us listen and learn about the world.

Copy Me!

Can you touch your ears and listen very quietly?

Part 3
The Sky & Weather Wonders

Let's look up and see the sunny,
rainy, windy, and starry sky!

Copy Me!

Can you point to the sun,
cloud, moon, and rainbow?

Why does the big yellow Sun shine down on us?

The Sun gives us light and warmth for the daytime. It helps make the world bright and cozy.

Copy Me!

Can you stretch your arms wide like the sunny sky?

Why do soft white clouds float way up high?

Clouds are made of tiny drops of water.
They float high in the sky and move along with the air.

Copy Me!

Can you puff your cheeks like a fluffy cloud?

Why does rain fall down from the clouds?

Rain gives water to thirsty flowers,
trees, and grass.
It helps many living things grow.

Copy Me!

Can you wiggle your
fingers down like raindrops?

Why does a sparkly rainbow pop out after the rain?

A rainbow appears when sunlight shines through raindrops. The light makes many pretty colors.

Copy Me!

Can you point to all the rainbow colors?

Why does the Moon change its shape in the dark night?

The Moon looks different as it moves around Earth.
The Sun lights up different parts of it.

Copy Me!

Can you make a round moon shape with your arms?

Why is the sky blue in the daytime?

Sunlight spreads through the air,
and blue light shows up the most.
That helps the sky look blue to us.

Copy Me!

Can you point up
to the big blue sky?

Why does the wind blow and swirl around us?

Wind is moving air all around us.
We can feel it when it whooshes by.

Copy Me!

Can you blow soft air like the wind?

Why does snow fall down in soft white flakes?

When the air is very cold,
water can turn into snowy flakes.
Then the flakes fall softly from the sky.

Copy Me!

Can you twirl around
like a snowflake?

Why do little stars twinkle in the night sky?

Stars look like they twinkle because their light travels through the moving air. They sparkle high above us.

Copy Me!

Can you wiggle your fingers like twinkling stars?

Part 4
Toys, Play & Everyday Magic

Let's explore bouncy toys, colorful art, and playful everyday magic!

Copy Me!

Can you point to the ball, bubbles, crayons, wheel, and magnet?

Why do bouncy balls bounce high when you drop them?

A bouncy ball squishes when it lands.

Then it springs back up again!

Copy Me!

Can you bend down and bounce up like a ball?

Why do soap bubbles float and pop so fast?

Soap bubbles are tiny soap walls filled with air. They float for a little while, then pop!

Copy Me!

Can you puff your cheeks and blow like bubbles?

Why do crayons leave pretty colors on your paper?

Crayons leave soft waxy color when we draw. That is how pictures get bright and pretty.

Copy Me!

Can you pretend to draw a big rainbow?

Why do wheels roll so fast across the floor?

Wheels are round, so they turn
and roll smoothly.
That helps toys move across the floor.

Copy Me!

Can you make your hands
turn in big circles?

Why do magnet toys stick together like magic?

Magnets have a hidden pulling power.
They can pull some pieces close together.

Copy Me!

Can you bring your hands together like magnets?

Why do blocks stack up high?

Flat sides help blocks sit on top of each other. That helps us build tall towers.

Copy Me!

Can you stretch up tall like a block tower?

Why does a spinning top twirl around?

A spinning top twirls when we give it a little spin. It keeps turning until it slows down.

Copy Me!

Can you twirl around like a top?

Why do puzzle pieces fit together?

Puzzle pieces have special shapes that match. When the right shapes meet, they fit together.

Copy Me!

Can you make your hands fit together?

Why do teddy bears feel soft to hug?

Teddy bears are made with soft fabric
and fluffy filling.
That makes them cozy to hold.

Copy Me!

Can you give yourself a cozy teddy hug?

Part 5
Yummy Food & Kitchen Fun

Let's explore tasty food, kitchen surprises,
and yummy everyday fun!

Copy Me!

Can you point to the ice cream, banana, apple, soup, and water?

Why does ice cream melt so fast on a hot day?

Heat makes ice cream warm and melty.
That turns it from frozen to drippy and soft.

Copy Me!

Can you pretend to lick
a yummy ice cream cone?

Why do bananas wear yellow jackets?

The peel keeps the banana safe and fresh.
Its yellow jacket protects the sweet fruit inside.

Copy Me!
Can you pretend to peel a banana?

Why do apples grow up high on big trees?

Apple trees grow fruit on their branches.
Up high, apples can drink sunshine and grow sweet.

Copy Me!

Can you reach up high like an apple tree?

Why does soup bubble and get hot on the stove?

Heat warms the soup and makes tiny bubbles.
That is how soup gets hot and steamy.

Copy Me!

Can you blow gently like you are cooling soup?

Why do we drink cool water when we run and play?

Water gives our bodies a cool drink.
It helps us feel fresh when we get hot and thirsty.

Copy Me!

Can you pretend to take a big sip of water?

Why do strawberries have tiny seeds outside?

Strawberries carry many tiny seeds on the outside.
Each little seed helps make new strawberry plants.

Copy Me!

Can you point to the tiny seeds on the strawberry?

Why does bread turn crispy and toasty?

Heat makes bread warm and dry on the outside.
That is why toast feels crispy and smells yummy.

Copy Me!

Can you pretend to pop up like toast?

Why do carrots crunch when we bite them?

Carrots are firm and crunchy. When we bite them, they make a crisp little crunch.

Copy Me!

Can you pretend to crunch a crunchy carrot?

Why do cookies smell yummy in the oven?

Warm cookies send sweet smells into the air.
That yummy smell tells us they are baking.

Copy Me!
Can you sniff the air like you smell fresh cookies?

Part 6 Nature & Growing Things

Let's explore flowers, trees, seeds, grass, and all the happy things that grow!

Copy Me!

Can you point to the flower, tree, grass, and little sprout?

Why do flowers open in the morning?

Flowers open up when the sunshine wakes them.
They stretch wide to drink light
and say hello to the day.

Copy Me!

Can you stretch open wide like a flower?

Why do green leaves grow on tree branches?

Leaves grow on branches so they can catch sunshine. The leaves help the tree make food and stay strong.

Copy Me!

Can you lift your arms up high like tree branches?

Why do tiny seeds turn into big, tall plants?

Seeds grow when they get soil, water, and sunshine. Little sprouts wake up and grow taller and taller.

Copy Me!

Can you crouch down small, then grow up tall like a plant?

Why does green grass tickle our bare feet?

Grass has lots of little blades
that stand up tall and soft.
They can feel funny and tickly on our feet.

Copy Me!

Can you wiggle your toes
like the grass is tickling them?

Why do autumn leaves fall down to the ground?

When the weather gets cooler, trees let some leaves go. The tree rests, and new leaves can grow later.

Copy Me!

Can you flutter down slowly like a falling leaf?

Why do plants need water to grow?

Plants drink water through their roots.
Water helps them stay fresh,
strong, and growing.

Copy Me!

Can you pretend to pour water on a thirsty plant?

Why do roots stay under the soil?

Roots stay under the soil to hold the plant steady. Roots also drink water and help feed the plant.

Copy Me!

Can you squat down low like roots under the ground?

Why do little buds turn into flowers?

A flower starts as a little bud.
As it grows, the bud opens into
a bright beautiful flower.

Copy Me!

Can you make your hands closed
like a bud, then open them
like a flower?

Why do trees give us cool shade?

Tree leaves block some of the hot sunshine.
That makes a cool shady place under the tree.

Copy Me!

Can you spread your arms wide like a shady tree?

Let's Remember!

We asked so many BIG questions and found tiny answers!

You are amazing for wondering about the world every day!

Keep looking, listening, and asking every day.

The world is full of wonder, just for you!

What Was Your Favorite Question?

Think about your favorite page in this book.
Draw it here!

You Can Keep Wondering!

There are so many more questions in the world!

Keep looking, listening, and asking every day.

Thank You!

Thank you for reading this book
with your big person.
You make a wonderful team!

The End

Little hands asked.
Big hearts learned.
Let's keep wondering together!

You are amazing!

www.ingramcontent.com/pod-product-compliance
Lightning Source LLC
LaVergne TN
LVHW060043210926
848481LV00013B/3948
* 9 7 8 1 4 0 5 8 2 8 3 1 4 *